YOUR KNOWLEDGE HAS VALUE

Sven-David Müller

Eating Right with Hemochromatosis. A Diet Guide for Reducing Iron

GRIN Publishing

Bibliographic information published by the German National Library:

The German National Library lists this publication in the National Bibliography; detailed bibliographic data are available on the Internet at http://dnb.dnb.de .

Imprint:

Copyright © 2005 GRIN Verlag GmbH
Print and binding: Books on Demand GmbH, Norderstedt Germany
ISBN: 978-3-656-61387-9

GRIN - Your knowledge has value

Since its foundation in 1998, GRIN has specialized in publishing academic texts by students, college teachers and other academics as e-book and printed book. The website www.grin.com is an ideal platform for presenting term papers, final papers, scientific essays, dissertations and specialist books.

Visit us on the internet:

http://www.grin.com/

http://www.facebook.com/grincom

http://www.twitter.com/grin_com

Hemochromatosis from a medical-nutritionist and dietetic view

By Sven-David Müller, M.Sc.

Hemochromatosis: When the liver stores too much iron

Hemochromatosis is a rare disorder of the iron metabolism, which leads to abnormal deposits of iron in the liver and other organs. Alone in Germany, an estimated two to four hundred thousand people suffer from hemochromatosis.

Hence, the so-called iron overload is among the most common hereditary disorders. In the format of this short communication we will discuss, what the characteristics of hemochromatosis are and how it can be recognized at an early stage. The disorder is usually diagnosed in patients between 40 and 60 years old. Primary hemochromatosis has a hereditary cause, whereas the secondary form of iron overload occurs with blood disorders. Patients suffer from a particular form of *diabetes mellitus* and dark pigmentation of the skin (bronzing), as well as hepatic cirrhosis. Other clinical syndromes include hormonal imbalances, cardiomyopathy and other physiological changes. Patients show elevated serum levels of iron and increased concentrations of ferritin. Routine treatment consists of phlebotomies. Moreover, extreme challenges such as food items rich in iron must be avoided. An iron-reduced diet, however, cannot replace phlebotomies as a form of therapy. The daily intake of meat and sausage products is not to exceed 120 grams (or 4.2 ounces). As a matter of principle, offal and processed products thereof must be avoided and it is recommended to consume cheese rather than sausage products.

Hemochromatosis (iron overload), which was described for the first time in 1889, is caused by an autosomal recessive gene defect, afflicting men at a ten times higher rate than women. The disorder is characterized by an increase in the total iron content of the individual from 4-5 g in the normal range to up to 80 g, due to an elevated iron absorption rate in the intestine. The first clinical presentation of the disorder is seen after the age of 20 at the earliest, more commonly between 40 and 60 years old. Women most often are affected after menopause, hence after cessation of their monthly period. Untreated hemochromatosis can lead to extreme fatigue, joint problems, diabetes, abnormal skin pigmentation, cardiomyopathy, hormone disorders, hepatic cirrhosis, to even liver carcinoma. Treatment consists of phlebotomies in regular intervals in the fashion of bloodletting. With timely diagnosis and treatment, life expectancy and quality of life are hardly compromised.

Hemochromatosis is one of the most common hereditary disorders in Europe. It is characterized by an increased iron uptake from the intestine to the blood, which is the vehicle transporting and ultimately depositing the iron in various organs. Iron in the right amount is indispensable for the synthesis of i.e., the red blood cells, or hemoglobin. Iron in excess amounts in the blood causes considerable damage through subsequent deposition in various organs.

The normal range of iron contained in the human body is 4-5 g. With hemochromatosis, the total iron content of the body varies between 20 and 80 g in comparison. A laboratory-based parameter is a 60% increase in transferrin saturation.

The term is explained in the section on iron in the body. Without treatment, which consists of regularly scheduled phlebotomy (bloodletting), this disorder bears significant health risks, including a marked decline in quality of life and lifespan. The onset of clinical signs is rarely seen before 20, rather between 40 and 60 years of age. Hemochromatosis is an autosomal recessive genetic disorder, autosomal meaning that the relevant gene defect is not located on a sex chromosome and recessive meaning that a respective carrier expressing the defect on a single chromosome is not affected phenotypically. In order for a descendant to have a chance of exhibiting symptoms, both parents have to be carriers of the defect. The gene responsible for the defect is located on chromosome 6 and is known as the HFE gene, with H standing for hemo- and FE for the chemical symbol of iron, Fe. The majority of mutations target the HFE1 gene, though very rare and specific forms of the disorder are encoded by mutations to the HFE2 and HFE3 genes. Men are affected at a ten times higher rate than women, since women are quasi subjected to natural therapy through their monthly menstruation. In the beginning stages, there are usually no noticeable symptoms. The affected individuals often do not know about their defect. The earliest onset of symptoms occurs after 20 years of age, in the majority though, between 40 and 60. As mentioned before, women almost exclusively show symptoms after menopause. With a total iron content of less than 10-15 g at the lower end of the abnormal spectrum, no symptoms are expected, which is termed the latent form of hemochromatosis. A histological examination, i.e. at the microscopic tissue-level, of the hepatic tissue, however, shows iron overload of those cells at this stage already.

A further increase of the iron content leads to the following symptoms and medical conditions, respectively:

- Fatigue, general weakness, indisposition
- Diabetes (*diabetes mellitus*)
- Discolorations of the skin
- Loss of libido, impotency
- Abdominal pain
- Shortness of breath
- Joint problems
- Hepatic cirrhosis to liver carcinoma
- Heart disease

Early diagnosis is imperative for a successful therapy. Tissue damage due to iron overload, resulting in cardiac myopathy, joint damage, diabetes or hepatic cirrhosis normally cannot be reversed, despite of intense therapy. Treatment of hemochromatosis consists of the very old method of phlebotomy, which is bloodletting. Hereby, initially about 500 ml of blood is drawn once or twice a week. One or two phlebotomies of 500 ml each per week can remove 200 to 400 mg of iron and are usually tolerated by the patients without any problems. In order to monitor the success of treatment, iron indicators should be assessed on a regular basis. In case the therapy led to a normalization of total iron levels, 4-6 phlebotomies per year are required for the rest of a patient's life. These individuals do not qualify as blood donors, since their blood does not range within the required norm. Medication is rarely given.

Iron overload – Hemochromatosis

Iron overload or hemochromatosis is one of the most common hereditary metabolic disorders. Close to half a million people in Germany suffer from the condition and an estimated 5-10% of the Central European population can potentially transmit the genetic defect to their children. This gene defect – passed on by father and mother – results in an increased iron uptake at the nutritional level, due to a dysfunction in the small intestine and subsequently iron deposition in various organs (see graph). As a result, the following symptoms are observed in various manifestations and combinations:

- Fatigue, decline in productivity, depressive mood, difficulties concentrating,
- Cramping of the upper abdominal area, irregular heart beat, shortness of breath,
- Joint pain (especially knee, hip, finger, big toe),
- Declining libido, impotency, irregular or missed periods, gray-brown skin (possibly tanning)

Consequential damage with long-term iron poisoning in untreated patients could be hepatic cirrhosis, *diabetes mellitus*, hormonal disorders, weakness and arrhythmia of the heart, arthropathy, and heightened risk for liver carcinoma.

Diagnostic methods of the disorder include iron saturation and ferritin levels (ferritin as the storage protein) in the blood and with an increase in those levels genetic testing is warranted, which in 80-100% of cases confirms the typical mutation on the maternal and paternal alleles of chromosome 6. Ambiguous cases sometimes require the assessment of a liver biopsy. Therapy of the iron overload condition consists of regular phlebotomies for life, in order to deplete iron stores. If bloodletting is not indicated because of anemia or other reasons, the drug Desferoxamin (Desferal) is the treatment of choice. A diet low in iron can support the treatment, though it is not sufficient as the sole form of therapy. Early diagnosis and consistent phlebotomy therapy are vital in avoiding secondary conditions and having a life expectancy that is normal.

Eat and drink right with hemochromatosis

The iron overload disorder hemochromatosis is not among the diet-induced conditions. Nevertheless, hemochromatosis patients do benefit from a balanced diet, which should be low in iron. Moreover, some nutritional principles in context with phlebotomies should be considered. It seems unreasonable for hemochromatosis patients to ingest a diet that is rich in iron. This is particularly exacerbated in a situation when the body takes up iron very efficiently. Meat and sausage products are an example for this food category. People with iron overload disorder should absolutely avoid multivitamin/mineral supplements, since those usually also contain iron. In addition, the commonly added vitamin C promotes iron uptake from food. Also, caution should be taken not to consume food items enriched in iron. Any enrichment has to be indicated on the food label. Whereas vitamin C promotes iron uptake, tannin in black tea, calcium in milk products, dietary fiber like pectin or phytate in whole grain products impede iron uptake. As a precaution and to treat a developing hemochromatosis, a reduction of iron in the diet could be advisable. It is recommended that hemochromatosis patients follow an iron-reduced, healthy and diversified diet. Zinc supplements are beneficial, since they block iron uptake.

Patients who suffer from hemochromatosis should consume less iron-rich food items, i.e., not daily and in large quantities. An important consideration is that the human body can utilize iron of animal origin more efficiently than iron of plant origin. Therefore, "iron-bombs" of animal origin are marked with an asterisk (*) and plant-based foods despite of a high iron content are not. This means, the latter are allowed, as long as they are not consumed in combination with additional vitamin C or other substances considerably improving iron uptake.

Foods with high iron content (* avoid and never eat with copious amounts of vitamin C)

Chanterelles, dried	57.6 mg/100 g	120.2 kcal/100 g
Yeast	20.0 mg/100 g	224.4 kcal/100 g
Maggi seasoning	20.0 mg/100 g	224.4 kcal/100 g
Homemade blood sausage*	17.0 mg/100 g	343.9 kcal/100 g
Pig's liver, cooked*	15.4 mg/100 g	123.3 kcal/100 g
Wheat bran	12.9 mg/100 g	172.3 kcal/100 g
Soy protein, textured (textured vegetable protein, TVP)	12.5 mg/100 g	285.1 kcal/100 g
Cocoa powder	12.5 mg/100 g	342.5 kcal/100 g
Pumpkin seeds, fresh	12.5 mg/100 g	560.2 kcal/100 g
Soy flower (defatted), debittered	12.0 mg/100 g	196.7 kcal/100 g
Veal kidney, cooked*	11.3 mg/100 g	116.4 kcal/100 g
Filet blood sausage*	10.3 mg/100 g	247.1 kcal/100 g
Soy beans, roasted	10.0 mg/100 g	359.0 kcal/100 g
Sesame, fresh	10.0 mg/100 g	559.0 kcal/100 g
Pig's kidney, cooked*	9.8 mg/100 g	114.7 kcal/100 g
Pine nuts, fresh	9.2 mg/100 g	575.5 kcal/100 g
Roast chicken liver, cooked*	9.2 mg/100 g	146.7 kcal/100 g
Millet grains, peeled	9.0 mg/100 g	354.0 kcal/100 g
Millet flakes	9.0 mg/100 g	354.0 kcal/100 g
Millet whole grain	9.0 mg/100 g	330.8 kcal/100 g
Sorrel, fresh	8.5 mg/100 g	22.2 kcal/100 g
Flaxseed, fresh	8.2 mg/100 g	372.4 kcal/100 g
Wheat germ	7.9 mg/100 g	313.8 kcal/100 g
Soy beans, dried	7.8 mg/100 g	416.3 kcal/100 g
Veal liver, cooked*	7.6 mg/100 g	146.5 kcal/100 g
Scallops*	7.5 mg/100 g	77.0 kcal/100 g
Veal liverwurst*	7.4 mg/100 g	316.7 kcal/100 g
Chicken egg, yolk*	7.2 mg/100 g	348.7 kcal/100 g
Liverwurst, fine*	7.1 mg/100 g	328.4 kcal/100 g
Broad beans, dried	6.8 mg/100 g	326.0 kcal/100 g
Beef liver, cooked*	6.8 mg/100 g	147.0 kcal/100 g
Oyster, fresh*	6.7 mg/100 g	63.1 kcal/100 g
Soy meat with spices, dry product	6.7 mg/100 g	305.2 kcal/100 g
Oyster, freshly cooked*	6.7 mg/100 g	65.0 kcal/100 g
Liver pâté*	6.6 mg/100 g	299.5 kcal/100 g
Chanterelles, fresh	**6.5 mg/100 g**	**11.5 kcal/100 g**
Porcino mushrooms, dried	6.4 mg/100 g	148.9 kcal/100 g
Sunflower seeds, fresh	6.3 mg/100 g	574.8 kcal/100 g
Bakery products for diabetics	6.2 mg/100 g	351.8 kcal/100 g
Whole grain zwieback for diabetics	6.2 mg/100 g	351.8 kcal/100 g
Chickpeas, dried	5.9 mg/100 g	325.3 kcal/100 g
Oats, whole grains	5.8 mg/100 g	353.3 kcal/100 g
Pastry for diabetics	5.6 mg/100 g	414.4 kcal/100 g
Veggieburger, dry product	5.5 mg/100 g	298.0 kcal/100 g

Parsley, fresh herb	5.5 mg/100 g	52.6 kcal/100 g
Tomato concentrate	5.5 mg/100 g	175.2 kcal/100 g
Mussels, freshly cooked*	5.1 mg/100 g	68.6 kcal/100 g
Beef heart, cooked*	5.0 mg/100 g	102.5 kcal/100 g
Legumes, mature	5.0 mg/100 g	277.7 kcal/100 g
Foods with low or no iron		
Fruit yoghurt with sweetener	0.1 mg/100 g	64.3 kcal/100 g
Fruit sour milk with sweetener	0.1 mg/100 g	62.4 kcal/100 g
Butter	0.1 mg/100 g	741.2 kcal/100 g
Whey	0.1 mg/100 g	24.9 kcal/100 g
Bouillon with noodles or dumpling (R)	0.1 mg/100 g	10.0 kcal/100 g
Cream of asparagus soup (R)	0.1 mg/100 g	30.3 kcal/100 g
Herbal tea with sugar (drink)	0.1 mg/100 g	8.8 kcal/100 g
Cream cheese, double fat	0.1 mg/100 g	335.3 kcal/100 g
Cream cheese spread	0.1 mg/100 g	335.3 kcal/100 g
Cream cheese, fat	0.1 mg/100 g	281.3 kcal/100 g
Herbal tea (drink)	0.1 mg/100 g	0.7 kcal/100 g
Cod liver oil	0.1 mg/100 g	882.6 kcal/100 g
Dairy ice cream	0.1 mg/100 g	84.8 kcal/100 g
Coffee substitute with condensed milk and sugar (drink)	0.1 mg/100 g	13.9 kcal/100 g
Coffee substitute with sugar (drink)	0.1 mg/100 g	10.0 kcal/100 g
Peanut oil	0.1 mg/100 g	879.8 kcal/100 g
Cow's milk, skimmed and heated	0.1 mg/100 g	36.8 kcal/100 g
Coffee substitute with condensed milk (drink)	0.1 mg/100 g	6.2 kcal/100 g
Coffee substitute (drink)	0.1 mg/100 g	2.2 kcal/100 g
Non-dairy ice cream	0.1 mg/100 g	60.7 kcal/100 g
Cow's milk for consumption, skimmed	0.1 mg/100 g	36.1 kcal/100 g
Margarine for cooking	0.1 mg/100 g	709.8 kcal/100 g
Lard	0.1 mg/100 g	882.2 kcal/100 g
Margarine, plant-based, linoleic acid 30-50%	0.1 mg/100 g	709.8 kcal/100 g
Margarine, linoleic acid >50%	0.1 mg/100 g	709.1 kcal/100 g
Yoghurt, skimmed	0.1 mg/100 g	38.0 kcal/100 g
Kefir, skimmed	0.1 mg/100 g	37.8 kcal/100 g
Cow's milk, heated	0.1 mg/100 g	65.5 kcal/100 g
Cow's milk, partially skimmed, heated	0.1 mg/100 g	49.5 kcal/100 g
Yoghurt, full fat	0.1 mg/100 g	65.7 kcal/100 g
Yoghurt, partially skimmed	0.1 mg/100 g	46.1 kcal/100 g
Cow's milk, certified raw, full fat	0.1 mg/100 g	67.2 kcal/100 g
Sour milk	0.1 mg/100 g	63.6 kcal/100 g
Sour milk, skimmed	0.1 mg/100 g	34.2 kcal/100 g
Ambrosia dessert (R)	0.1 mg/100 g	57.9 kcal/100 g
Cow's milk for consumption, fat-reduced	0.1 mg/100 g	48.5 kcal/100 g
Cow's milk for consumption, full fat	0.1 mg/100 g	64.3 kcal/100 g
Sour milk 10% fat	0.1 mg/100 g	118.5 kcal/100 g
Kefir	0.1 mg/100 g	49.7 kcal/100 g

Sour milk, partially skimmed	0.1 mg/100 g	46.1 kcal/100 g
Creamy sour milk, full fat	0.0 mg/100 g	66.4 kcal/100 g
Butter, fat-reduced – half the milk fat	0.0 mg/100 g	382.6 kcal/100 g
Coke beverages, low-calorie	0.0 mg/100 g	3.6 kcal/100 g
Plant-based oils, linoleic acid 30-60%	0.0 mg/100 g	882.6 kcal/100 g
Sunflower oil	0.0 mg/100 g	882.6 kcal/100 g
Brandy, grain-based (spirits from grain)	0.0 mg/100 g	250.0 kcal/100 g
Coke beverages (caffeinated)	0.0 mg/100 g	60.7 kcal/100 g
Margarine, half fat, linoleic acid 30-60%	0.0 mg/100 g	361.9 kcal/100 g
Black tea with milk and sugar (drink)	0.0 mg/100 g	10.0 kcal/100 g
Black tea with sugar (drink)	0.0 mg/100 g	8.4 kcal/100 g
Black tea with milk (drink)	0.0 mg/100 g	2.4 kcal/100 g
Coconut oil, hydrogenated	0.0 mg/100 g	878.8 kcal/100 g
Soy oil	0.0 mg/100 g	871.9 kcal/100 g
Tea (drink)	0.0 mg/100 g	0.5 kcal/100 g
Beer, alcohol-free (<0.5wt% alcohol)	0.0 mg/100 g	25.6 kcal/100 g
Drinking water	0.0 mg/100 g	0.0 kcal/100 g
Beer, pilsner, lager	0.0 mg/100 g	42.3 kcal/100 g
Beer	0.0 mg/100 g	42.3 kcal/100 g
Schnapps	0.0 mg/100 g	185.0 kcal/100 g
Frying oil (mainly plant-based)	0.0 mg/100 g	884.1 kcal/100 g
Roast drippings (animal-based)	0.0 mg/100 g	878.1 kcal/100 g
Sweets for diabetics	0.0 mg/100 g	246.2 kcal/100 g
Cookies, low in protein and sodium, gluten-free	0.0 mg/100 g	235.4 kcal/100 g
Natural mineral water, not carbonated	0.0 mg/100 g	0.0 kcal/100 g
Wheat beer, top-fermented	0.0 mg/100 g	42.8 kcal/100 g
Wheat beer, export	0.0 mg/100 g	42.8 kcal/100 g
Beer, Starkbeer	0.0 mg/100 g	59.8 kcal/100 g
Baking powder	0.0 mg/100 g	155.6 kcal/100 g

Nutrition around bloodletting

In the course of a phlebotomy, the body looses about 200 to 250 mg of iron, and as a rule, the loss of 500 ml of blood is tied to this, which corresponds to roughly a tenth of an adult person's blood volume. In order to prevent circulatory problems due to the practice, hemochromatosis patients should never have it done on an empty stomach. It is rather advisable to drink at least half a liter of mineral-rich mineral water or apple juice mixed with carbonated mineral water and to eat something as well. During and after bloodletting the amount of fluid (500 ml) the body lost should be offset. In some doctor's offices this is done with an infusion, but it is quite appropriate to balance the loss of fluids with drinks. Isotonic drinks, like a mix of 1/3 apple juice and 2/3 mineral water, have proven particularly effective. The snack after bloodletting should be rich in carbohydrates, like for example a slice of bread with jam and a banana.

Ideal fluid intake

Just like a healthy person, a hemochromatosis patient should always drink plenty of fluids. The ideal quantity is 2 liters per day, whereas in the summer or with strenuous physical activity the fluid requirement may increase to 2.5 to 3 liters a day. Mineral water is a very good option, while coffee and black tea are considered stimulants. However, hemochromatosis patients can take advantage of the iron-blocking properties of black tea due to the tannins it contains. Therefore, it is recommended to drink a cup of black tea with milk with every meal, especially one that is rich in iron. In order to prevent overweight, the consumption of sugary drinks should be minimal and instead replaced with light options containing sweeteners. There is no carcinogenic effect of sweeteners by the way, or other negative effects when consumed in reasonable quantities.

Alcohol damages the liver and is not recommended with hemochromatosis

Alcoholic beverages are a bad choice for people who suffer from hemochromatosis, because the disorder targets the liver, leading to a compromised liver function to possibly liver cirrhosis. The international literature on the subject makes the recommendation that alcohol in small amounts – moderately – is allowed, if there is no evidence of liver damage. If liver damage is evident, alcohol is strictly prohibited in any form, including alcohol-containing chocolate truffles, chocolate, cough elixirs, and alcohol-free beer. Obviously, alcohol is not appropriate during and after bloodletting either.

An important trace element: Iron

Iron is a vital trace element. In comparison to most of the other minerals, the organism can store iron very effectively. Nevertheless, a healthy person should take up sufficient iron daily with a balanced, healthy diet. Iron plays an important role in the formation of blood, evident in the presentation of anemia as a result of marked iron deficiency. The iron content of food is only one factor of many contributing to an iron deficiency. The availability of iron within the organism is rather determined by losses through bleeding (menstruation), as well as an increased requirement due to growth and pregnancy. Moreover, the absorption rate of iron in the intestine is another factor that needs to be addressed with the development of an iron deficiency.

With regard to iron absorption, it has been demonstrated that complete meals are more advantageous than individual food items. Additionally, the form of iron presented is essential. Iron found in whole grain products and vegetables in large quantities is absorbed at a lower rate than iron from a meat source. Though, the supplementation with foods rich in vitamin C (fruits, vegetables) can considerably increase this absorption rate. The fruit added to muesli thus not only enhances the taste, but also increases the absorption rate of iron contained in the cereal flakes. Black tea contains many tannins capable of binding iron. Therefore, the absorption of iron is impaired, when black tea is ingested with a meal (same applies to coffee). Phytic acid, contained in whole grain products (especially fresh kernels or bran) in large quantities, binds iron in the intestine. Iron availability is depressed due to the formation of salts from food sources (phosphate, calcium). Likewise, a high content of dietary fiber (for example pectin) decreases the absorption rate of iron [3]. Based on a varied diet, between 14 and 60 mg of iron are ingested. A relatively small quantity of the iron contained in food is actually absorbed by the body. The absorption rate of iron varies between 1.4 and 30%. During the onset of hemochromatosis, the body constantly absorbs up to 20% of the iron from food sources [3]. As a result of bloodletting, the absorption rate of iron is increasing again [3]. Hence, it makes sense to follow a diet reduced in iron. The availability of iron is particularly enhanced, if it is offered together with acids – like ascorbic acid (vitamin C), lactic acid (in yoghurt and other acidified milk products), or acids in fruits (fruit juices). On the other hand, tannins (see following table) impair the absorption of iron. Tannins are found in black tea, for example. Acid-rich Sauerkraut has a particularly high iron availability of 30% [2]. The trace element zinc may interfere with iron absorption [4]. For this reason, zinc supplement products are recommended (15 mg zinc histidine mornings and evenings).

Improve iron absorption	Reduce iron absorption
Vitamin C	Tannins
Foods from animal sources	Calcium
	Phosphate
	Phytate
	Dietary fiber

Healthy nutrition
People who suffer from hemochromatosis should opt for a healthy and balanced nutrition. It is recommended to follow a diet, which includes limited quantities of iron-rich foods and enhancers of iron absorption, but plenty of iron absorption reducers. A healthy nutrition comprises a variety of foods of plant origin, like fruits, vegetables, salads, whole grain products, legumes, potatoes, rice, and pasta. These foods all contain relatively little iron, but a number of iron absorption reducers. Moreover, iron from plant-based food sources has a relatively lower availability as compared to that from foods of animal origin, thus hardly affecting your iron balance. A healthy nutrition is relatively low in fats, which are of plant origin, like vegetable oil or margarine.

In order to adequately compensate for protein lost with bloodletting, hemochromatosis patients should preferably choose milk products over meat and sausage products. Likewise, fish is a good and healthy source of protein. When it comes to eating three or five meals, this is entirely left to your convenience. A particular diet is not healthier for the reason of having five smaller instead of three bigger meals. Your contact person in all things "healthy nutrition and diet" is a dietician or nutritionist based in hospitals or with many health insurances. These professionals are trained in a three-year college program to provide assistance in those matters. Ecotrophologists (home economics and nutrition scientists) focus on the scientific aspects of nutrition.

Nutrition rules with hemochromatosis at a glance:

Avoid foods rich in iron: blood-containing foods like meat, sausage, liver, and kidneys (including sausage products with those ingredients – it is recommended to eat cheese instead of sausage and fish instead of meat).

- *No iron-enriched foods*
- *No mineral supplements containing iron*

No foods rich in vitamin C in combination with a meal high in iron
Drink black tea with every meal, in particular with foods rich in iron
Drink plenty

- *No alcoholic beverages*
- *Avoid alcohol also in prepared foods*

No bloodletting on an empty stomach
Drink plenty after bloodletting
Sufficient protein: at least 1 gram per kilogram daily

Dietary adjustments with health issues due to hemochromatosis

Hemochromatosis can cause *diabetes mellitus* and/or hepatic cirrhosis. Both conditions have their own set of dietary requirements, which do not contradict the needs of a hemochromatosis patient. Should you be diagnosed with diabetes or hepatic cirrhosis, seek the assistance of a dietician or nutritionist. Your doctor can also give advice.

Author:

Sven-David Müller, Master of Science in Applied Nutritional Medicine, federally accredited dietician and diabetes nutritional consultant with the German Diabetes Association (DDG), Haddamshäuser Weg 4a, 35096 Weimar an der Lahn, www.svendavidmueller.de, diaetmueller@web.de

Literature:

1) Fritz Heepe, Dietary Indications, published by Springer Verlag, 1990
2) Häufel/Ternes/Tunger/Zobel, Food-Encyclopedia, published by Behr's Verlag, 1993
3) Strohmeyer/Niederau in Nutritional Medicine by Biesalski, published by Thieme Verlag, 1999
4) Paolo M. Suter, Nutrition Checklist, published by Thieme Verlag, 2002

Book Recommendation:

Nutrition Guide Liver and Gall Bladder, published by Schlütersche Verlagsgesellschaft
Cover image: pixabay.com